What's Happening To Our Climate?

Messner Books by Malcolm E. Weiss

WHAT'S HAPPENING TO OUR CLIMATE?
STORMS—FROM THE INSIDE OUT
MAN EXPLORES THE SEA

What's Happening To Our Climate?

by Malcolm E. Weiss

Illustrated by Paul Plumer

Julian Messner New York

Published by Julian Messner, a Simon & Schuster Division of Gulf & Western Corporation, Simon & Schuster Building, 1230 Avenue of the Americas, New York, N.Y. 10020.

Second printing, 1979

Manufactured in the United States of America

Design by Phil Jaget

Library of Congress Cataloging in Publication Data

Weiss, Malcolm E.
What's happening to our climate?

Includes index.
SUMMARY: Discusses the changes in Earth's climate over the years and explains both natural and man-made reasons for them.
1. Climatic changes—Juvenile literature.
[1. Climatic changes. 2. Climatology]
I. Plumer, Paul, 1936– II. Title.
QC981.8.C5W44 551.6 78-15684
ISBN 0-671-32846-8

For Lee M. Hoffman
Editor and Friend

CONTENTS

Chapter

1. A Change in the Wind 11

2. What is Climate? 19

3. What Makes Climates Change? 33

4. Around the World—20,000 Years Ago 47

5. The Long Seasons 53

6. Smokestacks and Volcanoes 62

7. What Can We Do? 74

Glossary 84

Index 87

WHAT'S HAPPENING TO OUR CLIMATE?

Chapter 1.

A Change in the Wind

It was the year 1000 A.D.

The corn grew and ripened in the fields. It grew then as it grows now in the same land. Today we call that land Iowa.

Iowa is an Indian name, and the corn farmers of 1000 A.D. were Indians. We do not know what they called the land, but they had many villages.

They built their villages from Iowa west to the foot of the Rocky Mountains. In the thick forests, they hunted deer. Where the grass grew tall in the prairies, they planted their corn, beans and squash.

The weather changed with the seasons. Yet year after year, the *pattern* of the seasons stayed the same. Each season had its own kind of weather. The corn and other crops were harvested in the crisp cool days of fall. Some was saved to be eaten through the winter months when snow lay thick on the fields. In the spring, the snow melted, the ground softened and new crops were planted.

Year after year, the summer winds blew steadily from the Gulf of Mexico in the south. The wind carried rain and warmth from the sun-heated gulf to the crops and to the forests where the deer lived.

For two hundred years, these patterns of weather stayed the same. And then, very slowly, they began to change.

The south winds of summer blew less strongly and less often. More and more, as the years passed, the summer winds blew steady from the west.

The west wind was a dry wind. It brought sunny days and starry nights—too many of them. Without the rains, the soil dried out, the corn died and much of the forest, too. So did large numbers of deer, and other animals.

This was a *drought*, and it lasted two hundred years. In the far west, the foothills of the Rockies became *deserts*. The Indians left their villages and moved eastward.

Around the world, other weather patterns changed, too. In 1000 A.D., the Vikings were settling the southern part of Greenland, where the shores and the ocean were free of ice most of the year. Their small sailing ships could travel safely between Europe and Greenland. On the southwest shore, it was warm enough to graze cattle and to farm.

Here, too, as in Iowa, the weather began to change, although the change was very slow. The cold winter winds from the north began to come earlier in the autumn, and stayed later in the spring. Each summer, less snow melted.

As the cold crept down the coast from the north, the grass died and the cattle could not graze. More hundreds of years passed, and the

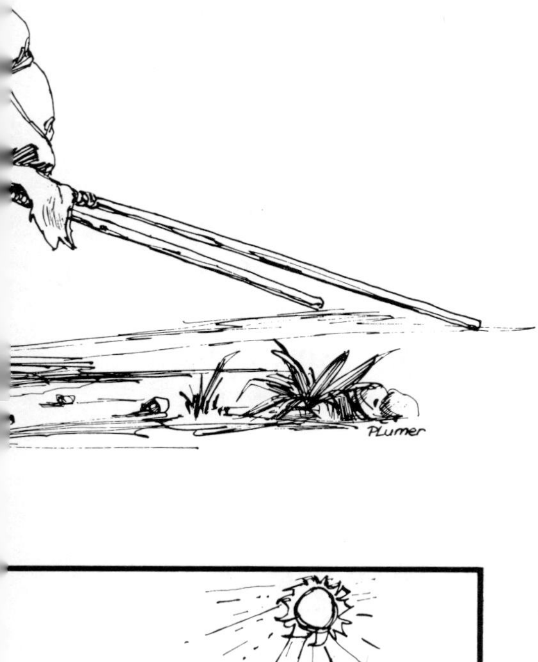

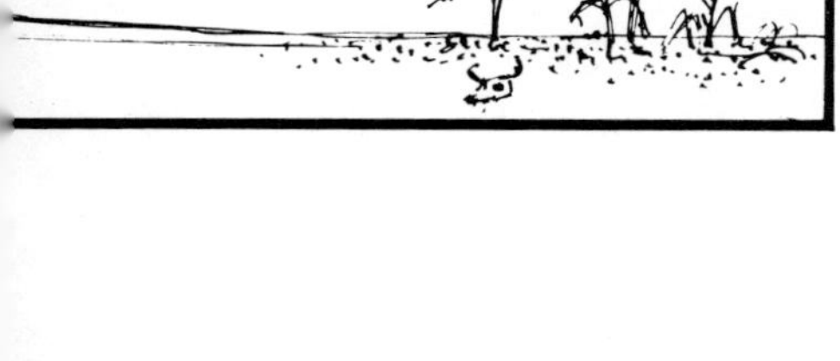

A 200-year-long drought dried up the forests and farmlands. Over the years, the Indians moved eastward, where rain was more plentiful.

farm buildings and lands were buried under slow-moving rivers of ice—*glaciers*. These glaciers reached the ocean, where big chunks of ice began to break off into the water—icebergs.

From about 1550 to 1850, the cold spread still further south. Scientists call that time the *Little Ice Age*. They gave it that name to contrast it with a great *Ice Age* that happened many thousands of years ago. That great Ice Age began about 100,000 years ago, and went on for about 90,000 years. It was much colder and lasted much longer than this little one of 1550 to 1850.

But even this Little Ice Age was colder than the weather we have today. One of the coldest winters of that time was 1780–81, when George Washington's ragged and hungry troops in the American Revolution huddled around their campfires in Morristown, New Jersey.

The savage winter weather of that year seemed to help the British soldiers who lived in the warm houses of nearby New York City, which they had captured. Parts of New York Harbor froze so solidly that the Redcoats could move their cannons across the harbor to New Jersey, where they were ready for battle.

From 1850 to about 1940, the weather in the northern half of the world slowly warmed up. By the 1940s, the northern half of the world was about two degrees Fahrenheit (one degree Celsius) warmer than in 1850.

That may not seem like much. But a drop of two degrees all over the world would be enough to bring back those harsh winters to much of America, Europe and Asia. And since 1940, the temperature has begun dropping again, very, very slowly, but steadily. It has dropped about one degree since then.

Is this cooling off likely to go on? Are harsh winters likely to return? Or will the weather warm up once more?

These are questions about weather. But they are not the kinds of questions that scientists who study the weather and make weather forecasts try to answer. These scientists—*meteorologists*—try to predict the weather for the next few days. And they sometimes make general forecasts for a few months ahead of time. For example, at the start of winter, meteorologists may predict that it will be a much colder winter than usual.

The questions we have asked are questions

about the weather over many, many years. They are questions about climates and how they may change.

Climate is the weather at a place over a long period of time. *Climatologists*—scientists who study climate—have discovered how climates have changed over millions of years. This gives them some idea of how climates may change in the future. The answers to questions about such changes are far from certain. Even so, knowing about climate changes in the past can help us prepare for the way climates may change in the future. To see why this knowledge can help, let's take a closer look at how climatologists decide what the climate of a place is.

Chapter 2.

What Is Climate?

The whole idea of climate comes from the Greeks. So does the word "climate" itself.

About 2,500 years ago, the Greeks explored parts of Africa and Asia. They discovered the land of Scythia, to the north of the Black Sea in what is now the Soviet Union. Sailing south, the ancient Greeks reached Libya in Africa.

The Greeks thought their homeland had practically perfect weather—not too hot or too cold, not too wet or too dry. Scythia and Libya were something else again.

Like modern explorers, the ancient Greeks complained about the weather in other lands. Scythia was too cold and too wet. Libya was too hot and too dry. It was good to get home again!

But the Greeks did more than complain about the weather in other countries. They wondered why the weather seemed to be different from place to place. Why was it hot in Libya, warm in Greece and cold in Scythia?

One thing the Greeks noticed was that at noon in Libya, the sun was high in the sky. At noon in Greece, the sun was lower. In Scythia, it was lower still.

The Greeks thought this meant that the earth was curved. As a traveler moved northward from Libya, the land curved away from the sun more and more. The sun was lower in the sky, and the weather colder.

About 500 B.C., a Greek scientist, Parmenides (par-min-uh-deez), divided the world into five weather, or climate, zones. Near the *equator*, where the sun always climbed high in the sky, was the torrid (hot) zone. North and south of the torrid zone lay the two temperate zones, where the weather was neither too hot nor too cold. Around the North and South Poles lay the two frigid (cold) zones, where the sun was low in the sky.

The Greeks called these zones "klimata," from a Greek word meaning "slope." Each zone had its own weather pattern. The pattern depended on how far the zone sloped away from the sun.

The ancient Greeks had the right idea. Climate *does* depend a great deal on how much heat a place gets from the sun. And that depends on how far north or south of the equator the place is. The further you get from the equator, the more slanted the sun's rays are. The slanted rays can't warm up the earth as much as the direct rays.

An experiment with a flashlight will help you see why this is so. Shine the flashlight straight down on a sheet of paper. The flashlight makes a small circle of light on the paper. That is the way the sun shines on the earth at the equator.

Now tilt the flashlight so that its rays strike the paper at a slant. The light makes an oval shape on the paper. This is the way the sun shines on the earth north or south of the equator.

You can see that the oval of light is bigger than

At the equator, the sun's rays can shine straight down on the earth. This is the way the flashlight at right is shining on the paper. North or south of the equator, the sun's rays are more slanted as they strike the ground—like the light from the flashlight on the left.

When the flashlight is slanted more and more, the same amount of light is spread out over a larger and larger area. As the sun's rays get more slanted, the same amount of light and heat is spread out over a wider and wider area. So the sun's rays heat the land and water at the equator the most. Places north and south of the equator get less light and heat from the sun. The North and South Poles get the least.

the circle of light. But both oval and circle are made by the same flashlight. The light hasn't gotten any brighter or dimmer. That means that the oval has the same amount of light in it as the circle. Since the oval is bigger, the light in the oval must be spread out more thinly—just as you'd have to spread the same amount of peanut butter more thinly over a whole slice of bread than over half a slice.

In the same way, a slanted ray of sunlight spreads out more thinly over the earth's surface than a ray that shines straight down. Both rays carry the same amount of heat from the sun. But the heat carried by the slanted ray must spread out over a bigger area.

So places at the equator get the most heat from the sun. The hottest climates are near the equator. The North and South Poles get the least heat from the sun. The polar areas are always cold.

The climates in the temperate zones have more variety to them. They are not always hot or always cold. Each year they go through the regular change of the seasons—from winter to spring to summer to fall and back to winter again.

As we have seen, the ancient Greeks divided the world into five climate zones. But in those

FROM CLOUDS
FROM WATER
FROM DUST
FROM ICE AND SNOW
WARMS THE AIR
WARMS THE GROUND

Not all of the sun's heat and light warms the earth. Some is reflected back to space from ice, snow and water. Some is reflected back to space from clouds and dust in the air. Only a part of the energy from sunlight warms the air or the ground.

times, much of the world was unexplored. No records were kept of the weather at any place over a long period of time.

Today, climatologists divide the world into many more different climate zones. The United States alone is divided into some 11 zones.

Within each zone, the weather patterns are more or less the same year after year. For example, suppose you live around Miami, at the southern end of Florida. There the weather is warm. Even in the winter months, you would not expect the temperature to dip below 60° (16°C) very often. And you'd be surprised to see so much as a single snowflake.

On the other hand, suppose you lived in Fairbanks, Alaska. Then you wouldn't be surprised by snow in *any* month. Fairbanks can expect temperatures below freezing at any time of the year. So city officials keep antifreeze in the fire hydrants all year around.

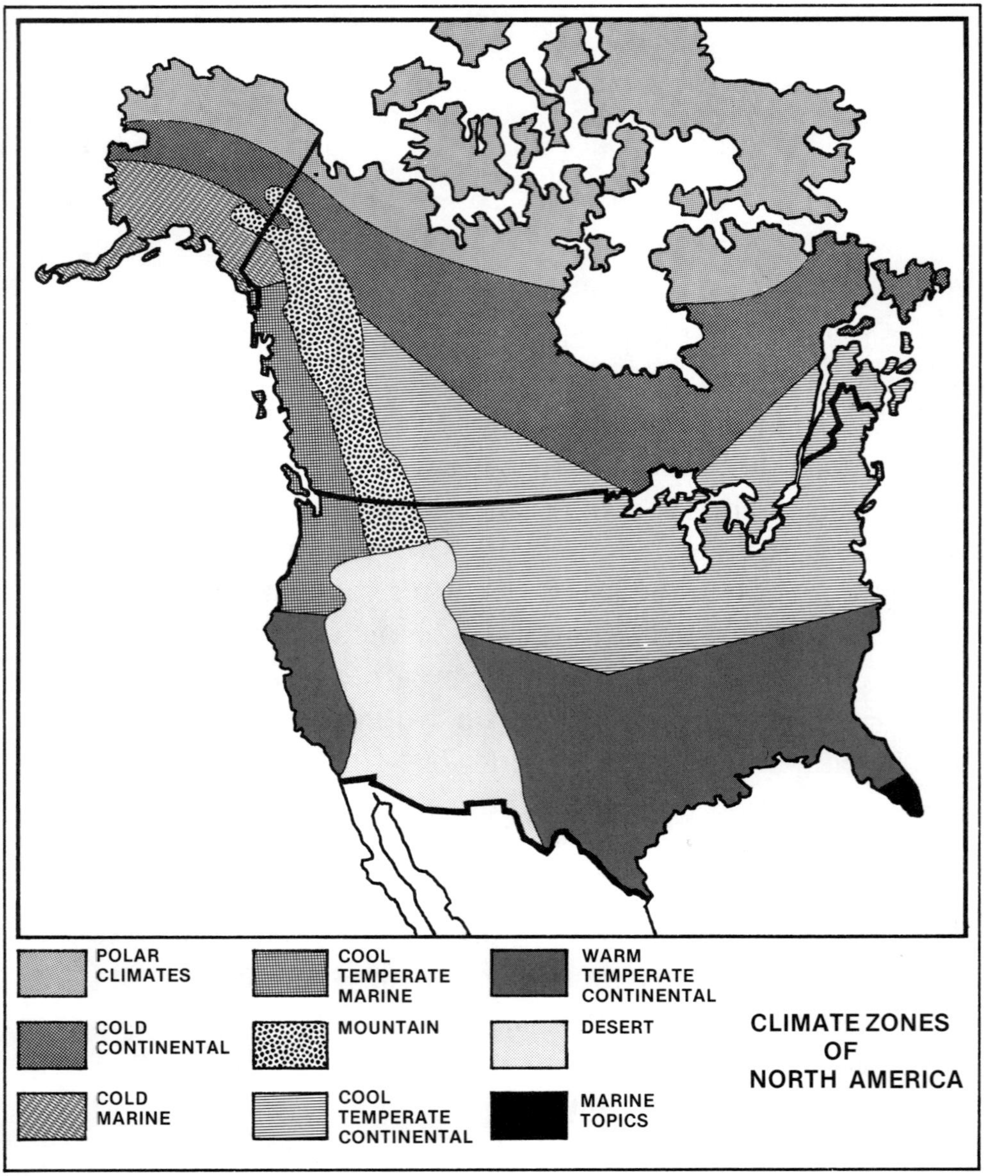
POLAR CLIMATES
COOL TEMPERATE MARINE
WARM TEMPERATE CONTINENTAL
COLD CONTINENTAL
MOUNTAIN
DESERT
COLD MARINE
COOL TEMPERATE CONTINENTAL
MARINE TOPICS
CLIMATE ZONES OF NORTH AMERICA

But in the winter of 1976–77, the weather did the unexpected.

Miami and other places in southern Florida had a little snow. It was the first time the weather service had ever recorded a snowfall in Miami.

In the middle of that same winter, people in Fairbanks had to cancel an out-of-doors ice hockey game. The ice was melting! Though it was January, the temperature was 50°. That's over 60 degrees warmer than the normal January temperature at Fairbanks, 12° (-11°C).

In fact, the winter of 1976–77 was unusual all over the United States. From the midwest to the east coast, it was the coldest winter on record. Cities like Chicago, Illinois, Pittsburgh, Pennsylvania, Jacksonville, Florida, and Nashville, Tennessee, had temperatures 20° (11°C) below normal. Buffalo, New York, had snow 60 days in a row.

In the far west, temperatures were near normal.

The climate of a place depends partly on how far north or south of the equator it is. But it depends on other things as well, as this map shows.

Mountain climates are cool or cold year round. This is because the air becomes colder at higher altitudes.

Places near the oceans have marine climates. The oceans warm up slowly during the summer and lose heat slowly during the winter. They help make marine climates cooler in summer and warmer in winter.

Places far from the oceans have continental climates. Their winters are colder and their summers are hotter. That's because land heats up and cools off much faster than water.

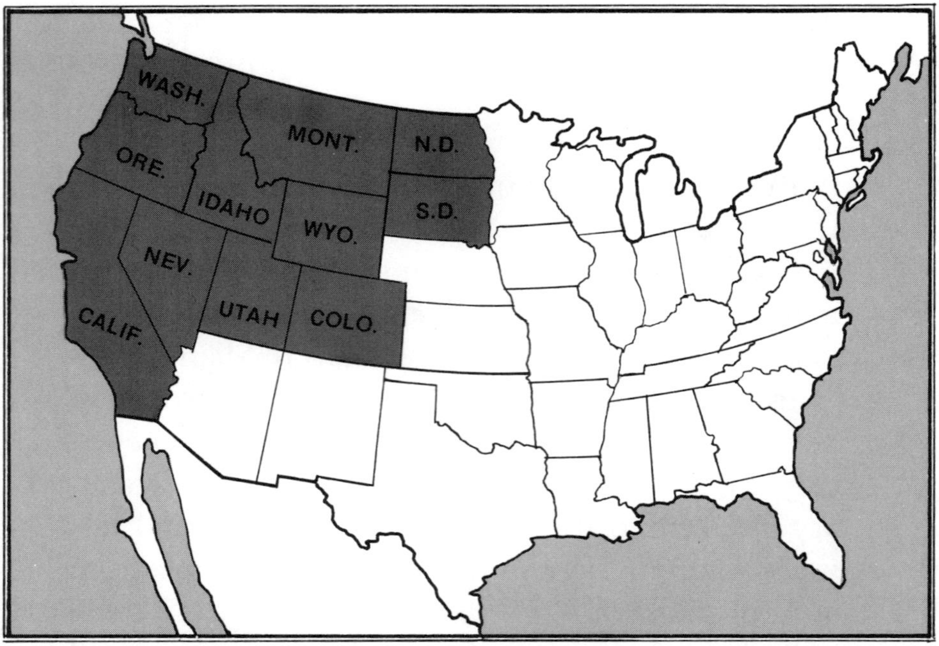

The shaded states had severe drought during the winter of 1976–77.

But 11 states there had a worse problem than cold—lack of water. Almost no snow or rain fell in this area during the entire winter.

Because of the drought, a lot of crops were lost. In parts of California, strict water-rationing laws were passed. Each person was allowed to use about as much water each day as you would need to take a five-minute shower. But that amount was all they were allowed for *every* purpose—washing, cooking and drinking.

All these weather changes were connected. They were all part of one huge weather pattern.

Normally, moist air from the Pacific Ocean travels eastward across the northern United States during the winter months. The winter storms, with their snow and rain, follow the same path. This air is cool, but it is warmer than the air further north over Canada.

In the winter of 1976–77, this pattern of air flow changed. The moist air from the Pacific Ocean turned north before it reached the west coast of the United States. Therefore, little rain or snow fell in the far west. Instead, the rain and snow and milder weather moved north with the wind over the west coast of Canada and Alaska. Alaska had a heat wave.

The air turned east over northern Canada. It traveled over hundreds of miles of ice fields, becoming bitterly cold.

Then it turned south and east. Like a gigantic icy river, the cold air poured over the midwestern United States. Blizzards and freezing weather rode on the winds all the way to Texas and the Gulf of Mexico. Texans shivered in the cold and remembered the old folk saying: "There's nothing between Texas and the North Pole but a barbed-wire

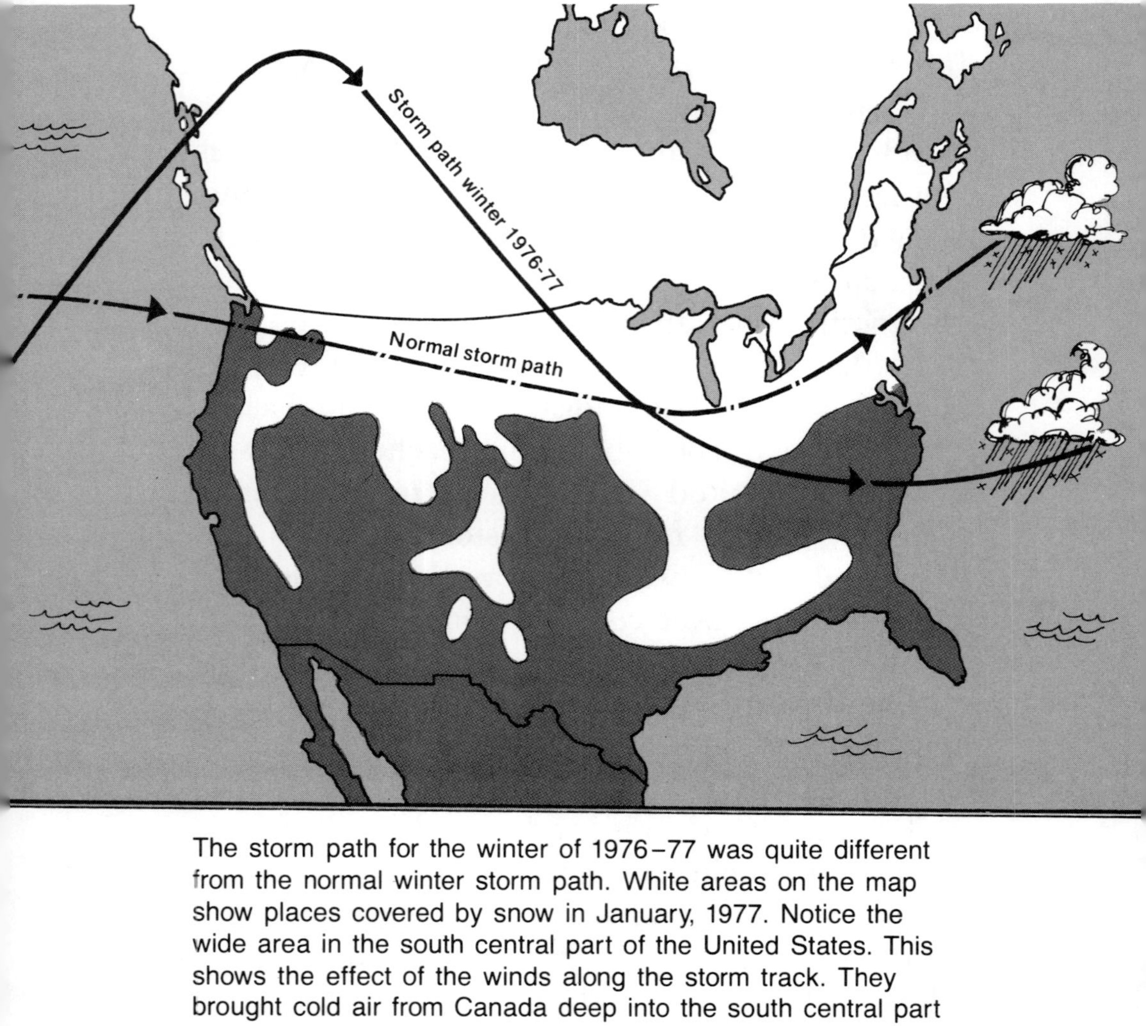

The storm path for the winter of 1976–77 was quite different from the normal winter storm path. White areas on the map show places covered by snow in January, 1977. Notice the wide area in the south central part of the United States. This shows the effect of the winds along the storm track. They brought cold air from Canada deep into the south central part of the country.

fence." And the whole eastern half of the country froze, too.

For nearly two months, this pattern of air flow stayed the same. Some scientists predicted that the same winter weather pattern would be repeated for the next few years.

Does this mean that climates in the United States are changing? Far from it. One year of strange weather, or even several years of strange weather in a row, doesn't mean a climate change.

But it does tell us something important about the way scientists measure climate. The weather records led scientists to expect one kind of winter weather pattern. The actual winter of 1976–77 was very different. Why? Are the records wrong?

The answer is no. Suppose we look at the weather records for the past 60 years in the United States. They show that only one other winter—the winter of 1917–18—was as bad as the winter of 1976–77. The other 58 winters were all milder.

Of course, the weather patterns did not repeat each other exactly over those 58 milder winters. No two years are exactly alike. In some winters, the weather was a little colder than usual and in other winters it was a little warmer. That's why climatologists try to look at what the weather has been like over a long period of time—the longer the better. The more years they can look at, the clearer their picture of what the normal climate for a place is like.

This tells us what climatologists mean when they say the weather for a particular season has been normal. They mean that it has been pretty much like the past weather for that season over a good many years. We expect future weather patterns to follow normal weather patterns of the past. But there is no guarantee that they will.

And climates can change. When that happens, what is normal changes, too.

Chapter 3.

What Makes Climates Change?

Before we can answer that question, we have to look at what makes weather work.

The sentence, "*W*hat makes *w*eather *w*ork?" has three "w's" in it. That gives us an easy way of remembering the three main things that make weather work: warmth, wind, water. They are the three "w's" of weather.

WARMTH and WIND

Warmth is the first and most important "w."

As we have seen, places at the equator get the most warmth from the sun. Places at the North and South Poles get the least. It is this uneven heating of the earth that makes the winds blow.

Near the equator, the air is heated by the warm ground or water beneath it. The warm air rises, because warm air is lighter than cool air. A few miles above the ground, this warm air flows toward the poles. Some of it flows toward the North Pole. Some of it flows toward the South Pole.

Wind, the second "w," is air in motion. This moving air is a wind that blows from the equator toward the poles.

At the same time, cold air from around the poles moves toward the equator to replace the rising warm air. Because this cold air is heavier than the warm air it replaces, it travels near the ground.

Warmer air is always moving toward the poles from the equator. And colder air is always moving toward the equator from the poles. Without this constant movement, the poles would be much colder and the equator would be much hotter than they are.

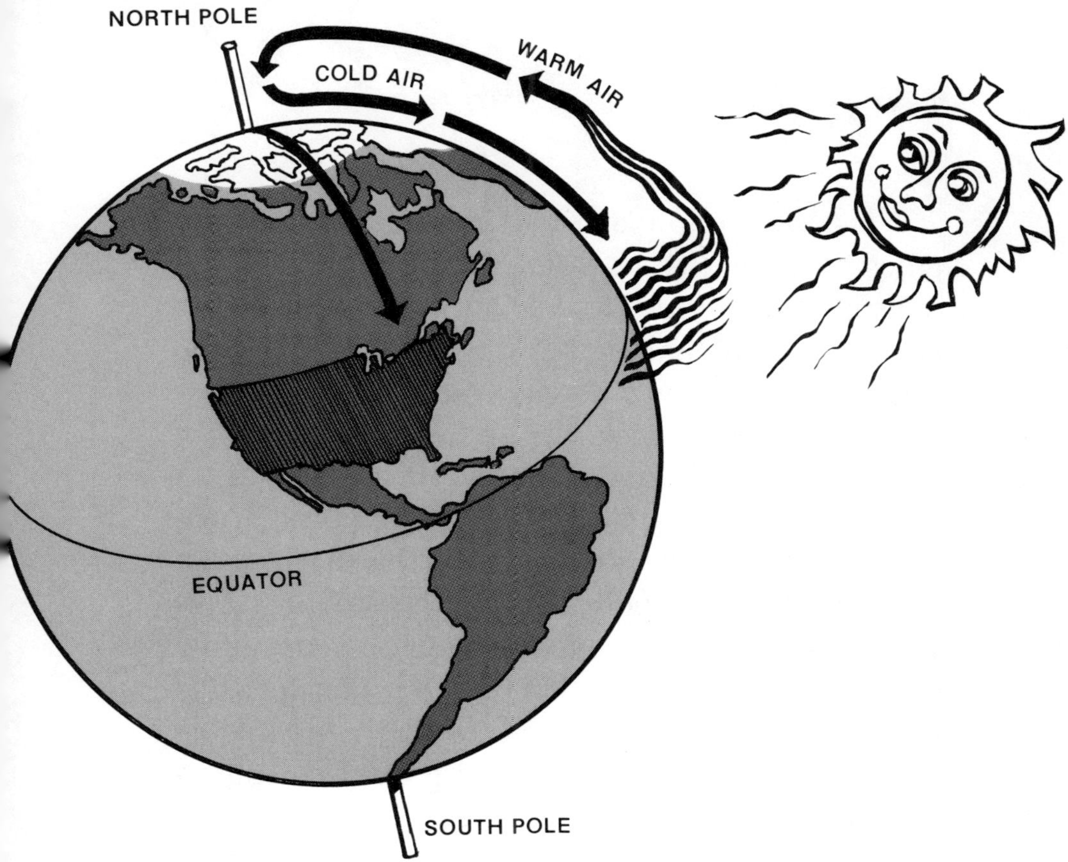

The sun heats the air most at the equator and least at the poles. Warm air at the equator rises and flows north and south. Only the northward flow is shown here. The southward flow makes the same pattern.

The warm air flows north toward the pole and cools. As the air cools, it becomes heavier and settles toward the ground. Near the ground, cool air flows back toward the equator.

So the uneven heating of the earth by the sun makes the air flow north and south. But the turning of the earth twists the paths of the moving air. (See illustration on page 37.)

This same circulation of warm and cold air takes place in a room that has a radiator or hot-air register at one end. Hold a bit of fine thread over the radiator or hot-air register. You can see the thread stream upward with the warm air. Now hold the thread high in the air at the other end of the room. Pushed by the warm air above, it should point *away* from the radiator. Next, hold the thread near the floor. Pushed by the cool air, the thread should point *toward* the radiator. This will work especially well if the room opens onto a cool hallway on the side opposite the radiator. The radiator is the hot "equator" of the room, while the hall is the cold "pole." The greater the difference in temperature between the warm and cold sides of the room, the more strongly the winds will blow.

If the earth were just like this room, the way the winds blow would be very simple—and very uncomfortable. In the United States, for example, there would be a steady cold wind blowing from the north over the ground. A few miles up, there would be a nice warm wind blowing from the south. We would look up toward it longingly, and shiver in the cold.

But the earth isn't like the room. It's spinning from west to east, and that makes a difference.

Suppose you try to roll a marble from the edge of a record player turntable to the center. When

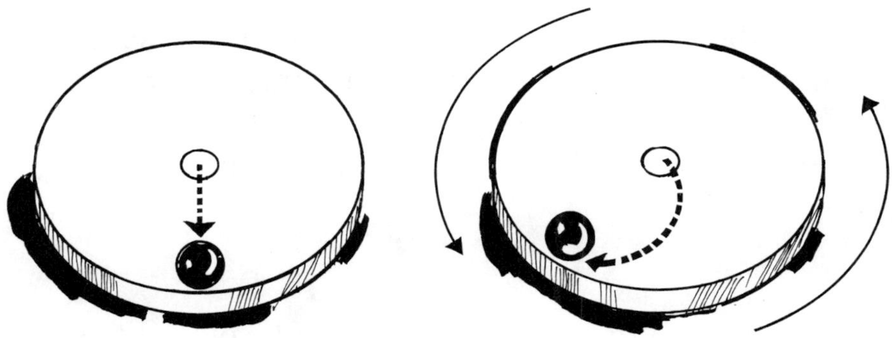

The turntable at left is not turning. A marble will roll in a straight line from the center of the turntable to the edge. Now, suppose the turntable is turning. Try aiming a marble at the center at the same point on the edge. The marble will twist to the left and miss the point. If you try rolling the marble from the edge to the center of a moving turntable, what happens?

In the same way, the turning of the earth causes the winds to twist and turn in certain ways. North of the equator, a wind blowing from the north twists toward the west. A wind blowing from the south twists from the east. South of the equator, it's just the other way around. A wind from the north twists east. A wind blowing from the south twists west.

the turntable is not moving, the marble will roll in a straight line. But when the turntable is spinning, the marble will twist as it rolls away from the center. If you try to roll it from the center to the edge, it will twist in the opposite direction.

In the same way, winds moving north and south over the earth twist and turn. This makes patterns of winds called *prevailing winds*—winds that blow from the same direction most of the time.

WATER

These winds bring warmth and cold. They move the storms that give us rain and snow. And rain and snow are both made of water—the third "w" making weather.

Without water, lands turn into deserts. But water is very important to weather and climate in another way, too. Water is a great storehouse of warmth.

Suppose on a cold winter day, you fill a glass with water. Put that glass and an empty one outside for about fifteen minutes. When you bring them in, which feels warmer?

The water-filled glass will feel warmer. The "empty" glass, of course, is filled with air. In the room, both air and water were at the same temperature. Outside, both the air and the glass lost their heat much more rapidly than the water. So the glass with water does not let the cold air in, and the water keeps the glass warmer longer.

For the same reason, the world's oceans help to store up warmth from the sun. In the winter, the oceans cool off more slowly than the land. So a city near the ocean stays warmer in the winter than an inland city. For example, the average January temperature in San Francisco, which is on the Pacific Ocean, is about 50° Fahrenheit (10°C). The average January temperature in Kansas City—which is as far north as San Francisco, but in the middle of the United States—is about 30° Fahrenheit (−1°C).

The oceans also warm up more slowly in the summer. In July, the temperature in San Francisco averages about 70° Fahrenheit (21°C). In Kansas City, it averages about 85° Fahrenheit (29°C).

The nearness of the Pacific Ocean gives San Francisco a mild climate. And oceans help to make climates in other ways, too.

Like the land, the oceans are heated unevenly by the sun. Ocean waters near the equator get far more heat than those near the poles. The upper

Uneven heating of the oceans produces powerful flows of water—the ocean currents. Prevailing winds and the turning earth help control the motions of these currents. Most current systems form a kind of giant wheel of moving water in the sea.

Ocean currents have a great effect on climate. The Gulf Stream, a warm current, helps warm much of northern Europe. Labrador, in Canada, is no further north than England, but it is much colder because of the cold Labrador Current. The cold Peru Current has helped turn a long strip of Chile and Peru into a desert. The West Wind Drift circles the icy Antarctic continent, cooling seas and lands to the north.

OCEAN CURRENTS

LABRADOR CURRENT
GULF STREAM
PERU CURRENT

layers of the oceans are heated by the sun, but the deeper layers are not.

Uneven heating of the air moves the winds. Uneven heating of the oceans creates mighty ocean currents. They flow through the ocean as winds blow through the air.

Warm water is lighter than cold water. The warm water near the equator flows toward the poles at the ocean surface. Cold water from the poles flows along the ocean floor toward the equator.

The turning of the earth twists these north and south currents, just as it twists the winds blowing north and south. And near the surface of the ocean, the prevailing winds also change the direction of the currents. About halfway between the poles and the equator, the winds blow toward the east. Near the equator, the winds blow toward the west.

The result is an ocean current flowing with the winds—westward near the equator, and eastward farther north or south. When these currents reach the edges of continents, they are forced to turn aside. So most current systems form a kind of giant wheel of moving water within the ocean.

For example, the Gulf Stream, a warm ocean current, flows past the east coast of the United States from Florida to North Carolina. The Gulf Stream flows on the ocean surface. Off North Carolina, the Gulf Stream turns northeastward toward Europe. One branch of the current flows around the British Isles; another flows farther north along the coast of Norway. At the same time, cold water is flowing south along the floor of the Atlantic Ocean just west of Europe. This current turns eastward near the equator, crosses the Atlantic, and rises as a warm current in the Gulf of Mexico. Leaving the Gulf of Mexico, it curves north around Florida, becoming part of the Gulf Stream.

The Gulf Stream is like an enormous river of warm water in the sea. The current carries as much water as a thousand Mississippi Rivers put together. This huge flow of warm water gives the British Isles a mild climate.

Edinburgh, in Scotland, is as far north as Labrador in northeastern Canada. The normal January temperature on the Labrador coast is 0° Fahrenheit (−18°C). But—thanks to the Gulf Stream—the normal January temperature in Edinburgh is about

42° Fahrenheit (6°C).

The effect of the Gulf Stream on the climate along the coast of Norway is even more impressive. Narvik is one of Norway's major ports. Narvik is actually in the Arctic, over 1,000 miles farther north than Labrador. Yet Narvik's normal January temperature is around 23° Fahrenheit (−5°C). The port's waters are ice-free all year round.

There's another reason why Labrador is so much colder than the British Isles or Norway. Sweeping down from the icy Arctic Ocean, a cold current, the Labrador Current, flows south along the Canadian coast.

Another cold current has helped turn a part of the west coast of South America into a desert. For 2,000 miles, the Peru Current flows north along the coast, from the middle of Chile to northern Peru. And from the middle of Chile to northern Peru stretch the unbroken wastelands of the Atacama Desert. It is one of the driest, deadliest places on earth. In some parts of this lifeless desert, no rain has fallen for 40 years.

Out to sea, in the Pacific Ocean, the air is full of moisture. Water from the ocean evaporates and goes into the air as water vapor, an invisible gas. Winds carry the moisture-laden air toward the coast.

But as the air crosses the cold Peru Current, it is chilled. Cold air cannot hold as much moisture as warm air. The water vapor in the chilled air turns back into liquid water. Before the winds can reach the desert, the water falls as rain. The Peru Current "squeezes" the air dry, as if it were a sponge. Only dry cool air crosses the desert from the sea.

Sometimes winds from the eastern Atlantic Ocean shores of South America blow over the desert. But they, too, are dry. Before these winds can get to the desert, they must cross the miles-high "wall" of the Andes Mountains. The air gets colder and colder as it rises over the Andes. On the eastern slopes of the mountains, the air loses its moisture as rain.

Warmth, wind and water work together to create weather and climate. A change in warmth can change the paths that winds and ocean currents take. When these changes last long enough, climates around the world change with them. Some places that got lots of rain or snow will get less and become drier. And some desert areas will get more rain or snow.

Such changes have happened many times in the past. The story of such changes is "written" in rocks on the land, and in the mud and sand at the bottom of the oceans.

Chapter 4.

Around the World—20,000 Years Ago

The place was where New York City stands today.

A howling blizzard piled up deep drifts of snow. Beneath the snow lay an ice sheet hundreds of yards thick. It was the southern edge of a gigantic sheet of ice that covered the land all the way to the North Pole. The storm was not unusual. Most of the year, it was stormy and windy. In the short, chilly summer, there were cold rains.

Summer and winter, the ice remained. It never melted. In fact, ice covered a large part of the northern half of the world.

It was another five thousand years before the ice sheet began to melt and shrink back to where it is today. The ice retreated very slowly, a fraction of an inch a year.

As the ice shrank, it revealed deep grooves in the rocks beneath it. The grooves had been cut in the rocks when the ice sheet moved over them. Here and there, the shrinking ice sheet carved long, narrow hills of sand and gravel into the land.

Some of these grooves and hills, rocks and lakes can be seen today. They are a part of a record kept by the earth itself—a record of climates some 20,000 years ago.

In Europe, such countries as Great Britain, Norway, Sweden, Finland and northern Soviet Union were under the ice. So was Denmark, and parts of Germany.

Even where Europe was ice-free, it was very cold. Tribes of hunters lived in central Europe then. Bones of the animals they hunted—reindeer, Arctic foxes and wolverines—can still be found. Today, these kinds of animals live much farther north in the Canadian Arctic and in Finland, where it is still very cold.

Today, sunbathers bask along the southern coasts of France and Italy. They swim in the warm waters of the Mediterranean Sea. Twenty thousand years ago, the same area was so cold that arctic foxes stalked through the snow, and reindeer galloped along the shores. Scientists have also found the shells of ancient clams that could only have lived in the Mediterranean Sea when it was very cold.

Twenty thousand years ago, in southern South America, and in southern Africa, much of the land was bare with scattered clumps of grasses and shrubs growing here and there. The plants that grew there tell us something about the climate then. It must have been cooler and drier. How do we know?

Scientists have found pollen from these ancient cold-climate plants. Pollen is a fine dust produced by male flowers. Blown by the wind, the pollen settles in the ground in layers. The oldest pollen is lowest down. Grains of pollen from each type of plant have their own special shape, which is unlike any other plant's pollen. So the pollen record

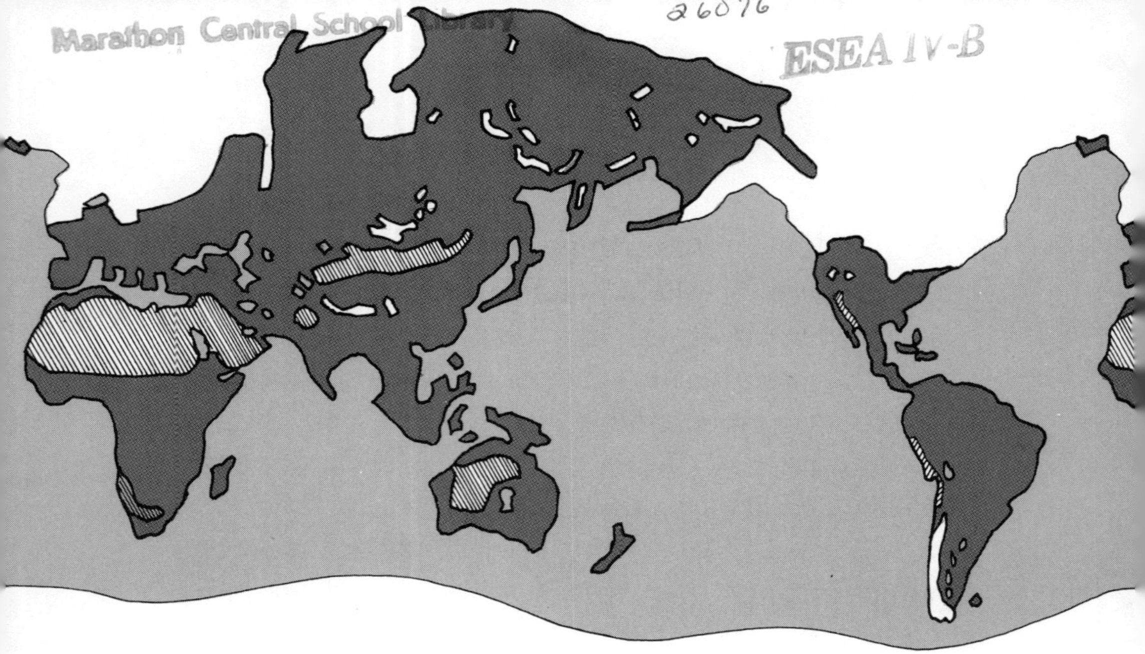

THE WORLD 20,000 YEARS AGO

Scientists have used many clues to put together this picture of what world climates were like 20,000 years ago. The world was much colder than it is now. Ice covered more of the land and seas of earth than it does today. In 20,000 years, the world's climates have changed very much.

tells the story of what plants lived in a place in times past.

The oceans were colder, too. Scientists have "taken the temperature" of ancient oceans by taking core samples of the ocean floor. To get the samples, they lower a hollow metal tube into the

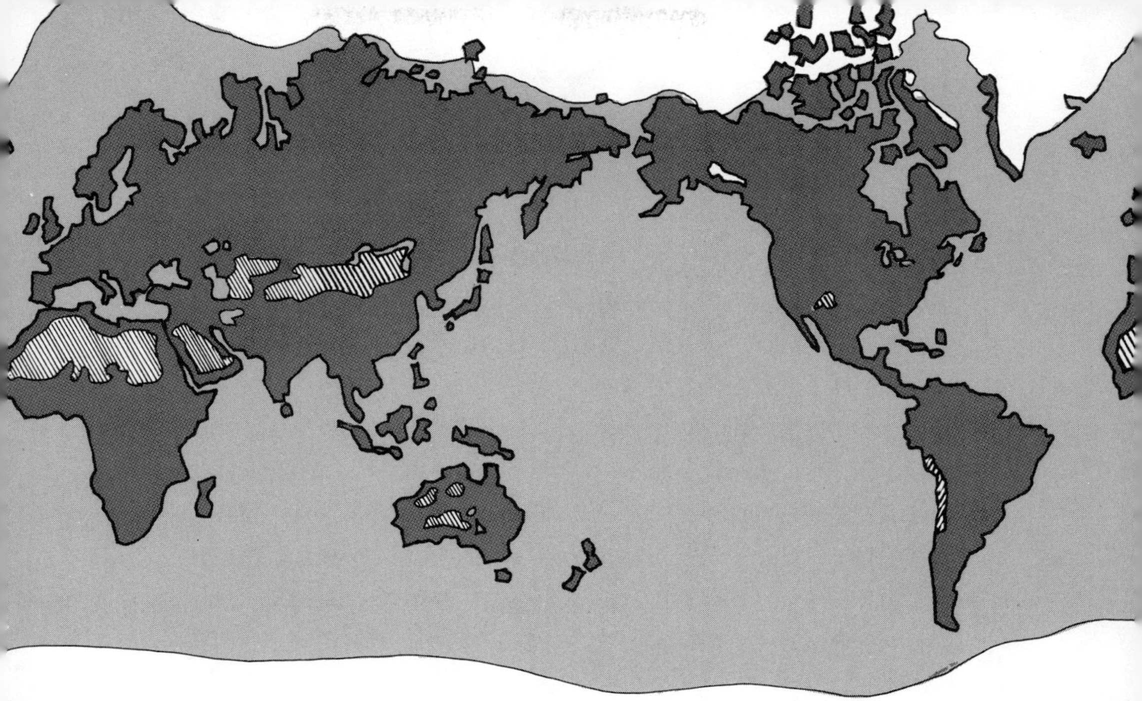

				THE WORLD TODAY
DESERTS	GROWING THINGS	OCEANS AND LAKES	ICE AND SNOW	

ocean floor. When the tube is hauled up, it is filled with dirt and sand from the bottom of the ocean.

In the core are the remains of many tiny sea animals. Like the pollen, the sea animals have settled in layers on the ocean floor. The ones that lived longest ago are furthest down. The sea animals of twenty thousand years ago needed cold water to live. They show that the oceans were colder than they are now.

This time of world-wide cold weather was the world's last great Ice Age. It was like a winter that lasted thousands of years instead of a few months.

In the past million years, there have been eight advances of ice sheets. Eight times, these ice sheets have crept south over North America and Europe. And eight times, the world has slowly warmed up again.

Only recently have scientists been able to put together this story of the Ice Ages. As we have seen, they have used the records kept by the earth itself—the traces of glaciers and stones, the remains of land animals and plants and tiny creatures of the sea.

It is a story of a slow, regular change from cold to warmth and back again. It is a change much too slow for anyone to see.

But we can compare it to changes that we can see—changes that are also regular but much faster—the changes in the seasons.

Chapter 5.

The Long Seasons

What makes the seasons? Why do they go through a complete change in just one year?

The earth spins like a top. But it doesn't spin like a top that is standing up straight. It spins like a top that is leaning over. Like the top, the earth is tilted.

We have summer when the part of the earth we live on is tilted *toward* the sun. The sun is highest in the sky, and day lasts longer than night.

Three months later, the earth has traveled a quarter of the way around the sun. The earth is still tilted in the same direction. But the earth's position in relation to the sun has changed. Now the earth is neither tilted toward the sun nor away from it. The sun is lower in the sky than it is in summer. Day and night are the same length. It is the beginning of fall.

In another three months, the earth has traveled halfway around the sun. The part of the earth we live on is tilted *away* from the sun. The sun is lowest in the sky, and night lasts longer than day. It is the beginning of winter.

After three more months, the earth is once again neither tilted toward the sun nor away from it. The sun is higher in the sky than it was in winter and day and night are equal. Spring has begun.

And after another three months, a year has passed. The earth is back to its position at the start of summer once again.

Not all places on earth have their seasons at the

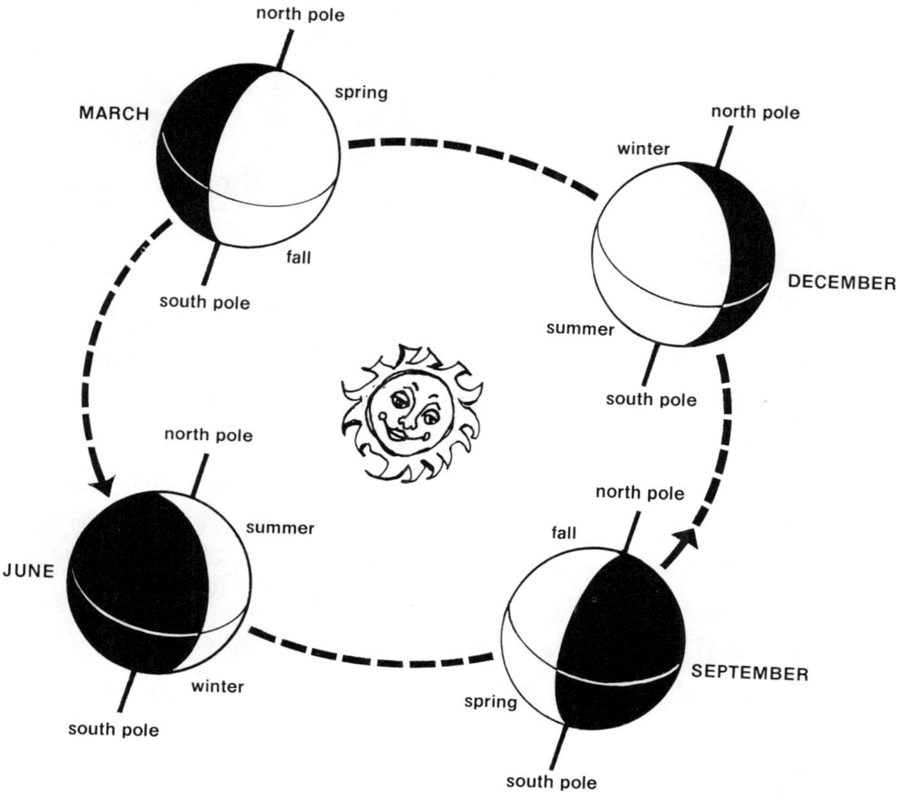

We have summer when the part of the earth we live on is tilted toward the sun. For people who live north of the equator, summer begins in June. The northern half of the earth is tilted toward the sun in June. But the southern half of the earth is tilted away from the sun at that time. So, south of the equator, winter begins in June.

Three months later, it is September. The northern half of the earth will soon begin to tilt away from the sun. There it is the beginning of fall. The southern half of the earth will soon begin to tilt toward the sun. There it is the start of spring.

In December, it is winter north of the equator and summer south of the equator. The southern half of the earth is tilted toward the sun. The northern half is tilted away from the sun. By the end of March, the northern half of the world is beginning to tilt toward the sun, and spring starts. The southern half of the world is beginning to tilt away from the sun. There it is the start of fall. In June, the earth has made a complete trip around the sun. A year has passed since the last June. Once more, it is summer north of the equator and winter south of the equator.

same time during the year. Look at the drawing. If you live north of the equator, *summer* begins in June. Around the North Pole, the sun never sets and there is no night. But if you live south of the equator, *winter* begins in June. Around the South Pole, the sun never rises and there is no day.

Half a year later, in December, it is winter north of the equator and summer south of the equator. Around the North Pole, the sun never rises and there is no day. Around the South Pole, the sun never sets and there is no night.

In the same way, when it is spring north of the equator, it is fall south of the equator. When it is fall north of the equator, it is spring south of the equator.

It takes one year for the earth to make a complete trip around the sun. So in one year, we go through the complete change of seasons and start all over again.

But the tilt of the earth is also changing. This is a very slow change, taking thousands of years.

Scientists say that nine thousand years ago, the earth leaned over more than it does today. Since then, it has been slowly moving into a more upright position. It will keep on getting more upright for about another eleven thousand years. Then it will start leaning over more again.

It's as if the earth were slowly nodding up and down at the sun. It is a very slow nod indeed. One up-and-down nod takes about forty thousand years!

This up-and-down nod changes the amount of warmth we get from the sun. The earth is tilted most at the "down" end of the nod. So each polar area gets the most heat from the sun during its summer. More of the ice around each pole melts. This helps make the earth warmer.

At the "up" end of the nod, each pole gets the least warmth from the sun during its summer. Year after year, more ice forms around each pole without melting. This helps make the earth colder.

Other slow regular changes in the motions of the earth also affect the amount of heat we get from the sun. Sometimes all these changes work together to make the earth colder. Winters get colder, and summers get cooler. Year after year, more ice may build up around the Arctic and Antarctic. That's because the summers there are then too cool to melt the ice that builds up during the winter.

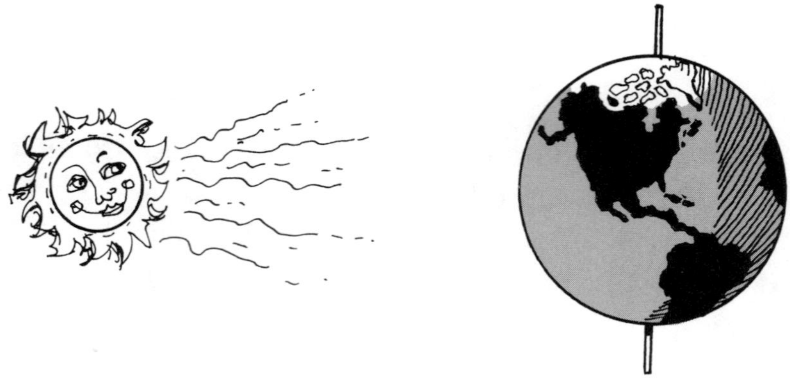

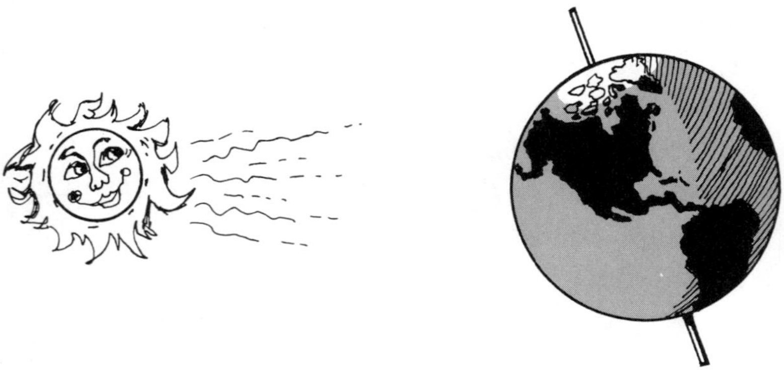

The tilt of the earth changes very slowly, as if the earth were slowly nodding up and down at the sun. One up-and-down nod takes 40,000 years.

This very slow change in tilt is one cause of the Ice Ages. At the "up" end of each nod (top drawing), each pole gets the least heat from the sun. Ice caps grow and spread around the poles.

At the "down" end of each nod (bottom drawing), each pole is tilted more toward the sun during summer. (The drawing shows summer for the northern half of the world.) The ice caps melt and shrink.

As more ice forms, it spreads over a larger area around the pole. And these large ice sheets themselves make the earth cool off faster.

How? You know that snow and ice glitter when the sun shines on them. Both snow and ice reflect a lot of the sun's light, like millions of tiny mirrors. The light—and heat—is reflected back into space and does not help heat the earth. So it is much harder for the sun to warm up land that is covered by ice or snow.

On the other hand, sometimes these changes in the earth's motions work together to make the earth warmer. Summers get hotter, and even winters get warmer. Year after year, more ice around the Arctic and Antarctic melts. More ice melts in the warmer summers. And the winters are too warm to form enough ice to replace all the melted ice.

As a result, the ice sheets grow smaller at their outer edges (toward the equator) and more and more land lies bare beneath the warmth of the sun. And that makes the earth warm up faster.

So changes in the motions of the earth helped cause the great Ice Ages, and the warmer times in between them.

As we have seen, the regular yearly change from cold to warm seasons and back again is caused by the earth's motion around the sun. The very slow change from an Ice Age to a warmer climate and back again is also regular. It is due to slow changes in the earth's motions. So we may think of these climate changes as the "long seasons" of earth.

Because of these changes, Dr. James D. Hays, a climatologist, says we are now in a time when the world is very slowly growing colder. He adds:

"This cooling trend should continue for about the next twenty thousand years. In that time, there is not much doubt that we will build up lots of ice in the northern half of the world."

That does not mean that an Ice Age is just around the corner. It will take thousands of years for these changes to cause an Ice Age.

However, Dr. Hays and many other scientists point out that it doesn't take a change to an Ice Age climate to cause trouble for the world. Even a little cooling off can make big problems for the world, especially in our supply of food.

Why? We've seen that such changes can shift the paths of storms. When that happens, there is less rain in some places that now have plenty of rain.

A slight cooling can also make the winters a little longer and the summers a little shorter. That means important food plants like wheat have a shorter time to grow. Less wheat may be produced. Yet the number of people in the world is large and growing larger. We must feed them.

Some scientists think the effects of a slight cooling trend are already showing up. The Soviet Union, the United States and Canada have had some poor weather in recent years. These countries grow most of the world's wheat.

But other scientists are not so sure. A few bad growing seasons tell us very little about how the climate is changing, they say. They agree that changes in the earth's motion can affect climate over thousands of years. But other changes may affect climate more quickly. Some of these changes are man-made.

Chapter 6.

Smokestacks and Volcanoes

Do you live near a city?

Do you live *east* of the city?

Then you can expect more rain than people living *west* of the city. You can also expect more snow. And you can expect more clouds, not to mention more smells from factories, and more fumes from cars, trucks and buses.

Why? In the United States, winds usually blow from west to east across the country. Let's follow the wind on its travels and see what happens.

The wind that blows through an American city may have been off the Pacific Ocean a few days before. It blew drops of salty seawater from the tops of the waves. The droplets dried up, leaving tiny crystals of salt floating in the air. The wind reached land, carrying the salt and the smell of the sea with it. Over land, the wind picked up more tiny solid particles—bits of dirt from the soil and perhaps dust from the smoke of forest fires.

Then the wind reached a city. It picked up more particles—man-made ones. It picked up dust from the smoke of burning oil, coal, gasoline and other fuels.

The solid particles in the air are important. Without them, no clouds would form. No rain would fall. No snow would swirl out of the sky.

Clouds, rain and snow are made of water. The water is in the air in the form of an invisible gas—water vapor. The vapor in the air turns back into thousands and thousands of water droplets so light that they can float in the air. Each droplet needs a particle around which it can form. If the

Man-made pollution—especially dust-laden smoke—can cause changes in local weather patterns. Country areas that lie just east of big cities have more cloudy days and more snow and rain than country areas that lie to the west of the cities. Prevailing winds, usually blowing from west to east, carry the smoke in that direction.

droplets grow big and heavy enough, they will fall as rain or snow.

Even "clean" unpolluted air has plenty of particles in it. The polluted air from a city has still more. Clouds form more easily in this air, and rain and snow fall from it more often.

That's why cities are cloudier, rainier and snowier than most of the countryside around them. But the area east of the city gets some of that snowier, rainier, cloudier weather, too. The wind brings it.

Man-made dust and dirt in the air has changed the weather in and around cities. But how will it change the climate?

We don't really know. Even the climatologists aren't sure.

One reason we don't know is that the dust and dirt can change the climate in different ways.

We've seen that this kind of air pollution makes more clouds. Too many clouds can keep the sun's heat from reaching the ground during the day. This might make the climate cooler.

The dust and dirt itself can stop some of the sun's heat from warming the ground. This, too, would make the climate cooler.

But at night, clouds are like a blanket that holds in the earth's heat. The sun warms up the earth during the day. At night, the earth slowly loses some of this heat into space. A nighttime blanket of clouds keeps the earth from losing as much heat as it does on a clear night. So nighttime clouds might make the climate warmer.

There's another reason we don't know how air pollution may change the climate: We haven't had enough time to find out.

Climate changes may take hundreds or even thousands of years to happen. But only in the past hundred years has man put enough pollution into the air to cause man-made weather changes.

Today, we can measure the effects of air pollution all around the world. For example, there is more carbon dioxide gas in the air now than there was fifty years ago. And there was more in the air fifty years ago than there was one hundred years ago.

Carbon dioxide is a normal part of the air. Animals breathe in oxygen and breathe out carbon dioxide. Fires use up oxygen and give off carbon dioxide.

Yet up to the past hundred years, the amount of

carbon dioxide in the air stayed about the same. That's because green plants used up the carbon dioxide given off by animals and fires.

Now we are adding carbon dioxide to the air faster than plants can use it up. Forest fires and farming have caused the loss of millions of acres of trees which once used carbon dioxide. The extra carbon dioxide in the air comes from the fuels we burn to heat our homes, run our factories and drive our engines.

Carbon dioxide keeps the earth from losing heat to space. The more carbon dioxide in the air, the more heat the air can hold.

Is the increasing amount of carbon dioxide in the air making the earth warm up? Once again, we are not sure.

From about 1880 to 1940, the earth warmed up slowly but steadily. During that same time, carbon dioxide in the air was increasing, too.

Since then, the amount of carbon dioxide has gone on increasing. But the temperature has been going down.

Why? The cooling may be due to other kinds of pollution in the air. We've seen that man-made dust and dirt might cause such cooling. Or it may not be due to any man-made cause at all.

Scientists just don't know.

However, there is another kind of "air pollution" that's been going on for a much longer time. It's been going on for millions of years. And scientists know more about the effects of this kind of air pollution.

This air pollution is caused by volcanoes.

Like giant natural smokestacks, volcanoes spew ashes, dust and gases into the air. In 1883, the volcano that formed the island of Krakatoa, in the Dutch East Indies (now Indonesia) erupted and then exploded. The roar of the explosion was heard three thousand miles away.

The huge amount of ash hurled into the sky blotted out the sun. It became pitch dark, although it was only ten o'clock in the morning. For three days, the sea around Krakatoa was in darkness, day and night. Dust fell on ships sixteen hundred miles away.

Fine dust from the explosion was thrown up at least thirty miles into the air. Today, the only man-made pollution that goes so high is from the trails of high flying jets and rockets and the clouds from atomic explosions.

An artist's view of the start of the eruption that destroyed Krakatoa in 1883. Sea water flooded into cracks in the volcano's side and boiled into steam. The enormous pressure of the steam pulverized the island. Bits of the island fell to earth over a thousand miles away. But most of Krakatoa was turned into fine dust and hurled some thirty miles into the air. That dust remained in the air for several years.

The air that high up is very cold and dry. Rain clouds never form to wash out the dust. Strong winds spread it quickly around the world.

The dust from Krakatoa stayed high in the air for several years. And for several years after the explosion, weather around the world was cooler than usual. Was the eruption responsible?

There were many volcanic eruptions in Iceland during the summer of 1783. The winter of 1783–84 was very cold. Were the eruptions responsible?

In April, 1815, another volcano in the Dutch East Indies blew up. A year later, *even the summer* was cold in New England. That was the year the farmers called "Eighteen-hundred-and-froze-to-death."

In Maine, as in all of New England, 1816 was a grim year. Every month had its share of freezing weather. There was plenty of snow in April. At the end of May, the ground was still too cold for planting. June came, and with it, snow and hail. By the end of June, farmers managed to plant some corn. But in early September, freezing weather killed it all.

Rumors spread that the climate was changing for the worse. Whole families and even whole villages set out west in search of a warmer, friendlier climate. Some fifteen thousand people moved away.

To find out if volcanic eruptions could cause climate changes, British climatologist Dr. Hubert H. Lamb checked records of volcano eruptions and weather going back hundreds of years. He found that the weather was almost always colder for several years after a major volcano erupts. The fine dust high in the air cuts off enough warmth from the sun to cause the cooling. After several years, the dust settles back to earth. The cooling stops.

We live in a time when big volcanic eruptions are rare. In the past sixty-five years, there has been only one. But suppose there were a big eruption every four years or so. The dust settling back from one eruption would be replaced by dust from a new eruption. The cooling effect would continue.

This is what happened from about 1550 to 1850. During this three-hundred-year period, there was a major eruption every four years. And as we have seen, the earth's climate was colder than usual. It was the time of the Little Ice Age.

We haven't had enough time to find out what man-made dust can do to the climate. But in a

Large volcanic explosions may produce layers of dust in the air that do not come down for years. The dust reflects more light and heat away from the earth into space, causing the climate to get cooler. A build-up of dust from man-made pollution may help cause the same kind of climate change.

way, the volcanoes have done an experiment for us. The experiment tells us that dust and ash thrown high in the air spreads round the world. It does not come down for years. And it can change our climate by shutting out some of the warmth of the sun.

The experiment is also a warning. Pollution can now be carried high into the air on the trails of jets, the flames of rockets, and the clouds of atomic explosions.

Enough of this pollution will change our climate. We do not know what the change will be like. We need to find out before the change takes place. Afterward, it will be too late to do anything about it.

Chapter 7.

What Can We Do?

How fast will the climate change?

How will it change?

Neither of these questions has a simple right answer. Let's look back briefly on some of the causes of climate change.

A slight change in warmth can change the winds. That can change the paths of storms around the world. Some places will get more rain and snow. Some will get less. All of these changes together mean a change in climate. Climate change begins with a change in the amount of warmth we get from the sun.

The change in warmth depends, in turn, on other things. It depends on the motions of the earth. It depends on how much snow and ice cover the ground. It depends on the number of volcanic eruptions. It depends on how much pollution we put into the air.

Sometimes these causes work together to make the climate change faster. The motions of the earth may cause the earth to grow colder. That may start more ice forming around the poles. And that may make the earth cool off still faster.

Sometimes these causes work against each other to make the climate change more slowly. Manmade dust and dirt in the air may tend to make the climate colder. At the same time, the extra carbon dioxide we add to the air may tend to make the climate warmer.

It's no wonder that even scientists can't agree on what's happening to our climate and why. And it's no wonder that the rest of us are even more puzzled.

After listening to all the arguments about climate, one newspaperman just gave up. "The best we can do," he wrote, "is to do nothing and hope that the climate experts are as wrong as the weatherman usually is."

Is that the best we can do? If the weatherman says a storm is coming, we don't just hope that he's wrong and forget about it. We put on a raincoat and take an umbrella. Or we dress warmly and wear boots.

The weatherman may be wrong. But we don't have to be certain about a danger to do something about it. Sometimes we can't wait to be certain before we act.

Scientists don't agree on why our climate will change, or in what way it will change. But they do agree that climate *will* change. Climates have always changed.

And they do agree that any change is likely to mean trouble for the world's food supply. Dr. J. Murray Mitchell Jr., an American climatologist, puts it this way:

"From the farmer's point of view, the climate's not going to get better. It can only get worse . . . We've been lucky because the weather in the major crop-growing areas of the United States has been so good for many years. To make plans as if the climate were going to continue to be so good for farming would be absolutely silly."

And at the end of 1976, the United States National Academy of Science issued a warning about climate change. It called the warm climate of most of this century "highly abnormal." It said that a change to colder conditions may already be starting. The scientists who wrote the report want the United States to prepare for such changes. Climatologists in other countries are telling their governments much the same thing.

Let's recall that climate simply tells us what the *normal* weather at a place is. Climatologists measure normal weather by looking at the weather records for a large number of years. Most of our measurements of normal weather use the records of the past fifty years or so.

However, many scientists think that the past fifty years have been unusually warm. If that is so, then these records are not a good measure of what normal weather is like. And they are not a good guide to the kind of weather patterns we can expect in the future. That is the point Dr. Mitchell and other climatologists are making.

What can we do about this problem? Scientists

are trying to find out more about why climates change. From 1978 to 1980, scientists will keep careful track of changes in warmth, winds and water around the world. Ships, planes, buoys, balloons, rockets and satellites will help in this worldwide weather watch.

We are learning how to make crops and land better able to withstand drought and cold. Corn is being used as a windbreak for sugar beets, for example. Because of this protection, the beet plants do not dry out so fast, and can go longer without water. A kind of soybean plant with crinkled leaves is being grown in some drought-prone areas. The crinkled leaves store large amounts of water that the plant can use when there is little rain.

The United States government is beginning to recognize that climate change is coming. In 1977, Secretary of Agriculture Robert Berglund ordered the Department of Agriculture to stop making forecasts on how much wheat will be grown "based on the assumption that we will have normal weather." That is something the Department has been assuming for many years.

There are other steps we need to take to prepare for climate change. We need to conserve our

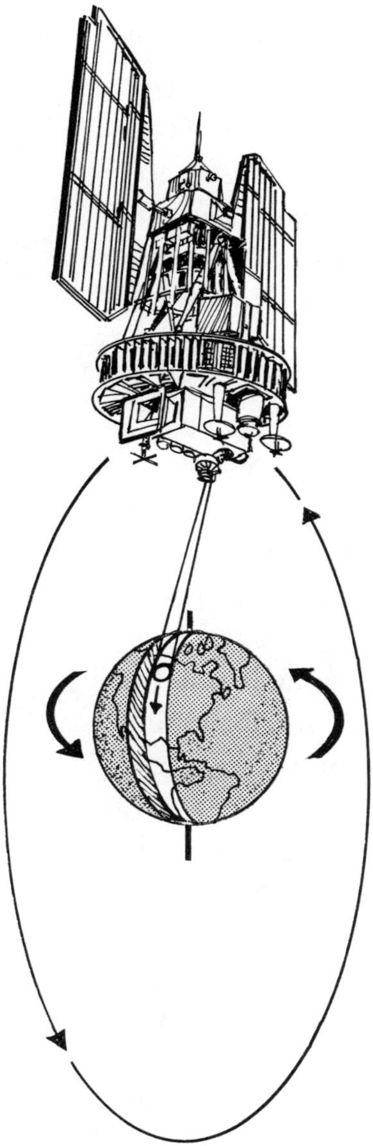

This kind of satellite can keep track of changing weather patterns around the world. The satellite circles the earth a number of times each day, passing over the North and South Poles. Each time, the satellite photographs a "slice" of the earth from the North Pole to the South Pole and back again. The white track on the globe shows half of one slice. The other half covers the opposite side of the globe and cannot be seen.

As the satellite circles earth, the earth is turning. When the satellite begins photographing the second slice, the earth has turned some distance from west to east. So the second slice shows a part of the earth a little to the west of the first slice. In twenty-four hours, the satellite can photograph the entire earth.

water supplies more carefully. People living in the far west must understand that drought is part of the normal weather pattern there. A hundred years ago, a scientist-explorer named John Wesley Powell warned people moving to the west that the drought would come and go in cycles. Today, scientists are repeating that warning.

And many scientists think the United States should store up food during the good growing years. They say this will help feed ourselves and the world when the bad years come. They point out that the United States has to take the lead in storing up grain, because this country grows by far the largest share of the world's food.

Planning ahead for climate change will not be easy. It will take time, hard work and lots of cooperation among farmers, the government and the rest of us. But we need to do it.

Other changes also force us to plan ahead. One is the energy crunch. Oil, our major source of energy, is scarce, and will be getting scarcer. We need to find new ways to heat our homes, power our factories, and run our cars. The choices we make may affect our climate.

For example, the United States has vast reserves

of coal. Coal can replace oil as an energy source for centuries to come. It can be burned as a fuel. It can be used to make natural gas and oil substitutes.

But all these uses of coal add carbon dioxide to the atmosphere, just as burning oil or gasoline does. From 1965 to 1977, the amount of carbon dioxide in the air increased by 5 per cent. In June 1979, scientists with the National Oceanic and Atmospheric Administration warned that carbon dioxide in the air is still increasing rapidly.

In the light of this information, should we choose coal as a major substitute for oil? It will take lots of money and time to get coal production in the United States going full blast again. Coal mines must be expanded, and new ones must be dug. Factories must be built to process the coal.

Once we get coal production going, we can't stop suddenly if we decide we've made a mistake. Many people's jobs will depend on the mines and the factories. We may become as "hooked" on coal as we are now on oil.

If we take this course, our major energy sources will continue adding carbon dioxide to the atmosphere at an increasing rate for years to come. In that case, scientists estimate that the amount of carbon dioxide in the air will double some time between the years 2025 and 2075. That may cause a rise in the average temperature around the world of 5.4°F.—3°C.

This would be warm enough to cause the polar icecaps to begin to melt. Sea levels would rise. The oceans would flood coastal lands where many large cities are located today.

Whether other changes would be good or bad from man's point of view is hard to predict. Plants use carbon dioxide to build new tissues, so plants might thrive on more carbon dioxide. On the other hand, the hotter weather might shift the American corn-growing belt north from the midwestern United States into central Canada. There, the soil is poor and much less corn could be grown.

A warmer earth would mean more rainfall. But heat also makes plants lose water faster. And the warm-up might change rain patterns so that places that get plenty of rain now would get very little.

Of course, we don't have to choose coal as an alternate energy source. We might decide instead to cut down on the use of energy sources that produce carbon dioxide. Then the extra carbon dioxide in the air will slowly be used up by plants.

What can we use to replace these energy sources? There are other sources that do not produce carbon dioxide. Among them are solar power, wind power, power from the tides and nuclear power. Each of these choices carries its own promise and its own risks. We have to think of them as we make our choices. We must learn to plan ahead more than we have in the past.

In the words of the National Academy of Science report on climate change, "We simply cannot afford to be unprepared."

We have learned to look *millions* of years into the past to see how climates have changed. We can plan a few years ahead, to be ready for the change that will come.

GLOSSARY

ANTARCTIC—The area around the South Pole. From two Greek words: "anti" meaning opposite, and "arctos" meaning bear. Near the South Pole, the Big Bear can never be seen. It never rises there.

ARCTIC—The area around the North Pole. It gets its name from "arctos," the Greek word for "bear." Near the North Pole, the constellation (sky picture) of the Big Bear circles round and round the sky without ever setting. (The Big Bear is the constellation that we call the Big Dipper.) *See* ANTARCTIC.

CLIMATE—The usual weather of a place over a long period of time.

CLIMATE ZONES—Areas of the earth that have similar climates.

CLIMATOLOGIST—A scientist who studies climates.

DESERT—A place where little or no rain falls, year after year.

DROUGHT—A long spell of weather with little or no rain.

EQUATOR—An imaginary circle around the

earth, halfway between the North Pole and the South Pole. *See* NORTH POLE.

GLACIER—A large mass of ice formed from snow. Glaciers form on high ground and move slowly down mountain sides or through valleys.

ICE AGES—Times in the past when climates around the world were much colder than they are today. These periods of cold lasted many thousands of years.

LITTLE ICE AGE—A time of unusually cold weather around the world that lasted from about 1550 to 1850.

METEOROLOGIST—A scientist who studies the weather. The word comes from the ancient Greek word "meteorologikos" meaning "about things high in the sky." To the ancient Greeks, the air above and around us was just a part of outer space. That was where the sun, the moon, the stars and the planets moved, and that was where weather happened, too.

NORTH POLE—The northern end of the earth's axis. An axis is the line around which a spinning object turns. Usually, the line is imaginary. The axis of a phonograph record would be a line straight through the hole in the center of the record. The earth's axis is an imaginary line from the North Pole through the center of the earth to the South Pole.

OCEAN CURRENTS—Large streams of water that flow through the oceans. Most currents are driven by the prevailing winds. Another cause of ocean currents is the uneven heating of the oceans by the sun. *See* PREVAILING WINDS.

PATTERN—A regular arrangement of forms or colors or events. In many countries, the four seasons are a part of the weather pattern.

PREVAILING WINDS—Winds that blow from the same direction most of the time.

SEASON—One of the four periods of the year: spring, summer, fall, winter. Each season has its own characteristic weather patterns.

SOUTH POLE—*See* NORTH POLE.

WEATHER—The state of the air at a certain place and time. For example, the air may be cloudy or clear, windy or calm, damp or dry, hot or cold. It may be raining, snowing, sleeting or hailing. *See* CLIMATE.

INDEX

A

Africa, 19, 49
air pollution, 63-66, 68, 73, 75
Alaska, 25, 27, 29
America, 17
American, 63, 76
American Revolution, 16
Andes Mountains, 45
animals, 13, 48-49, 51, 52, 66, 67
Antarctic, 57, 59
Arctic, 44, 57, 59
Arctic Ocean, 44
Asia, 17, 19
Atacama Desert, 44
Atlantic Ocean, 43, 45
atmosphere, 81-82

B

Berglund, Robert, 78
Black Sea, 19
British, 16
British Isles, 43, 44
Buffalo, New York, 27

C

California, 28, 39
Canada, 29, 43, 44, 61, 82
Canadian Arctic, 49
carbon dioxide, 66-67, 75, 81-83
Celsius. *See* temperature
Chicago, Illinois, 27
Chile, 44
climate, 18, 21, 23, 31, 32, 38, 39, 43, 44, 45, 48, 49, 60, 61, 65, 66, 70-71, 73, 74-83
climate zones, 20, 23, 25. *See also* individual names
climatologists, 18, 25, 31, 60, 65, 71, 76, 77
clouds, 63, 65, 66, 68, 70, 73
coal, 81
coastal lands, 82
corn, 11, 12, 13, 70, 78, 82
current systems, 42

D

Denmark, 48
Department of Agriculture, 78

deserts, 13, 38, 44, 45
dirt, man-made. *See* air pollution
drought, 13, 28, 78, 80
dust, man-made. *See* air pollution
Dutch East Indies (Indonesia), 68, 70

E

earth, 20, 21, 23, 34, 36, 37, 38, 42, 44, 48, 52, 53, 54, 56-57, 59, 60, 61, 65, 71, 75, 82
Edinburgh, Scotland, 43
energy sources, 80-83
equator, 20, 21, 23, 34, 36, 39, 42, 43, 56, 59
Europe, 13, 17, 43, 48, 52
experiment, 21-23, 36-39, 73

F

Fahrenheit. *See* temperature
Fairbanks, Alaska, 25, 27
farmers, 11, 70, 80
farming, 67
Finland, 48, 49
Florida, 25, 27, 43
food, 60-61, 76, 80
France, 49
frigid zone, 21
fuel, 63, 67, 81

G

gas, natural, 81
Germany, 48
glaciers, 16, 52
Great Britain, 48
Greeks, 19-21, 23
Greenland, 13
Gulf of Mexico, 12, 29
Gulf Stream, 43, 44

H

Hays, Dr. James D., 60

I

ice, 13, 16, 48, 57, 59, 60, 75
Ice Age, 16, 52, 59, 60
icebergs, 16
Iceland, 70

ice sheet, 47-48, 52, 59
Illinois, 27
Indians, 11, 12, 13
Indonesia. *See* Dutch East Indies
Iowa, 11, 12, 13
Italy, 49

J

Jacksonville, Florida, 27

K

Kansas City, Missouri, 39
"Klimata," 21
Krakatoa, 68, 70

L

Labrador, 43, 44
Labrador Current, 44
Lamb, Dr. Hubert H., 71
Libya, 19, 20
Little Ice Age, 16, 71

M

Maine, 70
Mediterranean Sea, 49
meteorologists, 17
Miami, Florida, 25, 27
Mississippi River, 43
Missouri, 39
Mitchell, Jr., Dr. J. Murray, 76, 77
Morristown, New Jersey, 16

N

Narvik, Norway, 44
Nashville, Tennessee, 27
National Academy of Science, 77, 83
National Oceanic and Atmospheric Administration, 81
New England, 70
New Jersey, 16
New York, 27
New York City, 16, 47
New York Harbor, 16
North America, 52
North Carolina, 43
North Pole, 21, 23, 29, 34, 39, 42, 47, 56, 75

Norway, 43, 44, 48
nuclear power, 83

O

ocean, 13, 16, 39, 42, 43, 44, 46, 50-51, 52, 82
ocean currents, 42, 43, 44, 45
oil, 80, 81
oxygen, 66

P

Pacific Ocean, 29, 39, 44, 63
Parmenides, 20
Pennsylvania, 27
Peru Current, 44, 45
Pittsburgh, Pennsylvania, 27
plants, 82
polar icecaps, 82
pollen, 49-50, 51
Powell, John Wesley, 80
prevailing winds, 38, 42

R

rain, 82
Redcoats, 16
Rocky Mountains, 12, 13

S

San Francisco, California, 39
satellites, 78
scientists, 16, 17, 18, 20, 31, 49, 50, 52, 56, 60, 61, 68, 75, 76, 77-78, 80
Scythia, 19, 20
seasons, 12, 23, 32, 52, 53-56, 60, 61
solar power, 83
South America, 44, 45, 49
South Pole, 21, 23, 34, 39, 42, 56, 75
Soviet Union, 19, 48, 61
sun, 12, 20, 21, 34, 39, 42, 54, 56, 57, 59, 60, 65, 68, 71, 73, 75
Sweden, 48

T

temperate zone, 20, 23
temperature, 17, 25, 27, 36, 39, 43, 44, 67
Tennessee, 27
Texas, 29

tides, 83
torrid zone, 21

U

United States, 25, 27, 29, 31, 36, 39, 43, 61, 63, 76-77, 80-83
United States government, 78, 80
United States National Academy of Science, 77, 80

V

Vikings, 13
volcanoes, 68-73, 75

W

warmth, 12, 33, 34, 38, 39, 45, 52, 59, 71, 73, 74, 75, 78, 82
Washington, George, 16
water, 16, 28, 33, 38, 42, 43, 44, 45, 78, 80, 82
water-rationing laws, 28
water vapor, 45, 63-65
weather, 12, 13, 16, 17, 18, 20, 21, 25, 27, 29, 31, 32, 33, 38, 45, 52, 65, 66, 70, 71, 76
weather forecasts, 17, 78
weatherman, 76
weather patterns, 12, 13, 21, 25, 29, 31, 32, 80
weather record, 46-52, 77
winds, 12, 13, 33, 34, 36, 38, 42, 45, 62, 63, 65, 70, 74, 78